Simulating Electromagnetic Wave Propagation Through Anisotropic Uniaxial Plate

For Normal Incidence With Coordinate-Free Approach

Surface 1	Air
	Uniaxial medium
Surface 2	
	Air

Ibrahim Ibrahim

DEDICATION

For those who love the topic of Electromagnetic Field Theory and its Applications.

CONTENTS

1 - INTRODUCTION

Electromagnetic wave propagation, with monochromatic plane wave properties, travelling through anisotropic uniaxial plate has been investigated and **statically simulated** in this thesis. By simulated, I refer to the simulation of the numerical results for the **theoretical analysis** (i.e., equations); this work is not concerned with any **numerical analysis** or methods of computation.

Reflection and Transmission of the propagated wave at the planes of interface, through the plate and eventually transferred outside the plate into the other side, were modeled. The model itself is unique in that it uses a coordinate-free approach that was, for the first time, introduced by H. C. Chen in his book, " *Theory of Electromagnetic Waves* ".

Detailed derivations and presentation of the equations are not included in this book. I will, in near future, produce another book for that purpose; but, in this work, you will certainly encounter all essential equations for our application.

Waves can be produced from various kinds of pulses, and for the simulation of our application I used RC_2 pulse. Matlab served as a perfect tool as well for generating and viewing the calculated results.

2 - THE APPLICATION

T he application that we have is illustrated below. You can see that the uniaxial plate is found in air and, therefore, sandwiched by it. On one hand, we have an air-to-uniaxial interface, which is Surface 1, and on the other hand, an uniaxial-to-air interface, which is Surface 2.

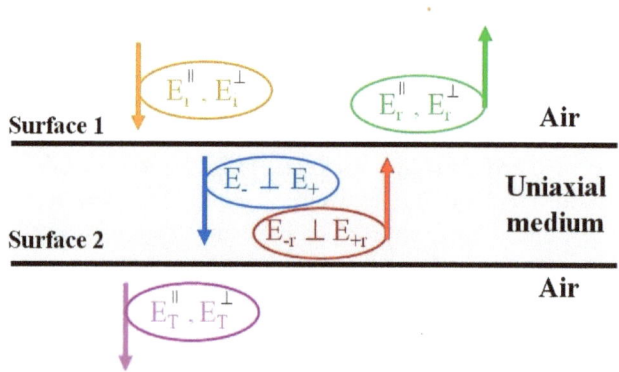

On each interface we have incident, reflected and transmitted waves. These waves can have perpendicular or parallel properties. Some have ordinary or extraordinary properties as the waves found inside the uniaxial plate medium. The incident waves on Surface 1 are $E_i^{\parallel}$ and $E_i^{\perp}$; the reflected waves are, $E_r^{\parallel}$ and $E_r^{\perp}$; the transmitted waves are E_- and E_+. On Surface 2, the incident waves are the same waves that transmitted the first interface, E_-

Simulating Electromagnetic Wave Propagation Through Anisotropic
Uniaxial Plate - For Normal Incidence With Coordinate-Free
Approach

and E_+; the reflected waves are, E_{-r} and E_{+r}; the transmitted waves are $E_T^{\parallel}$ and $E_T^{\perp}$. Notice in the illustration that the ordinary and extraordinary waves are perpendicular to one another.

The description of the application lies simply in the following sentence: Reflection and transmission of plane waves propagating in lossless, nonmagnetic and unbounded isotropic media through a plate of lossless, non magnetic and bounded homogeneous uniaxial media.

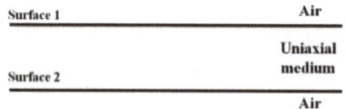

3 - ANALYTICAL MODEL

For deriving our mathematical model, I present first the General Method in the following set of equations:

General method

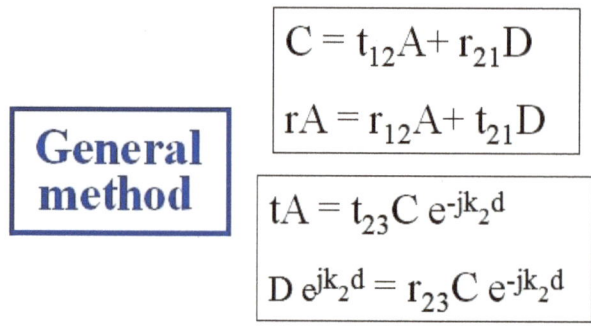

$$C = t_{12}A + r_{21}D$$

$$rA = r_{12}A + t_{21}D$$

$$tA = t_{23}C\, e^{-jk_2d}$$

$$D\, e^{jk_2d} = r_{23}C\, e^{-jk_2d}$$

These equations are illustrated in the figure below

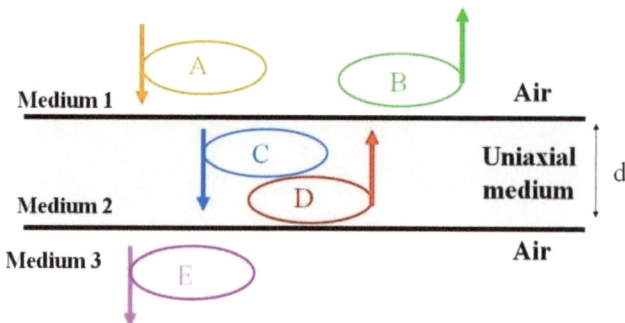

Notice here that I continue using the **Frequency Domain** representation as before for my model. You will see shortly in the next pages the illustration of the **Time Domain** model as well.

Another equivalent model to the, General Method, is, Airy's Summation Method; it has been laid down in 1833.

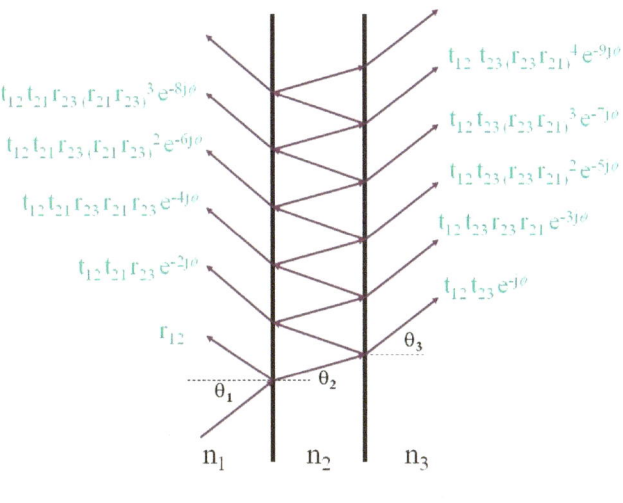

Airy's summation method-1833

Factor, e^{jkd}, in the equations of the General Method above is a function that sets and/or determines the position of the complex vector of **E**. Remember that you learned in Electromagnetic Field Theory about the, wave vector; this wave vector is represented here by **k**, which equals to $|k|\hat{k}$. The wave number here is, $|k|$, and $\hat{k}$ is called the, wave normal.

Solving the equations for the General Method model analytically, which I am writing in my next book, gives us the following relation

General method

$$r = r_{12} + \frac{t_{12}t_{21}r_{23}e^{-2j\phi}}{1 - r_{21}r_{23}e^{-2j\phi}}$$

Doing the same using Airy's Summation Method, gives us

Airy's summation method-1833

$$r = r_{12} + t_{12}t_{21}r_{23}e^{-2j\phi} + t_{12}t_{21}r_{23}r_{21}r_{23}e^{-4j\phi} + \ldots\ldots$$

These terms you are seeing in the equations above can be grouped, for better view, in the illustration below

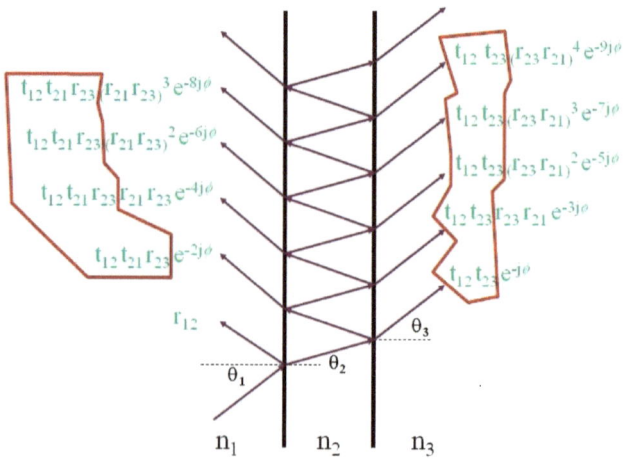

We are able to solve the equations using a simple
mathematical relation:

$$\frac{1}{(1-x)} = 1 + x + x^2 + x^3 + x^4 + \dots \quad \text{, for } |x| < 1$$

This relation qualifies both of the equations from the
General and Airy's Summation Methods to be equal
accordingly.

General method

$$r = r_{12} + \frac{t_{12}t_{21}r_{23}e^{-2j\phi}}{1 - r_{21}r_{23}e^{-2j\phi}}$$

Equal terms

**Airy's summation
method-1833**

$$r = r_{12} + t_{12}t_{21}r_{23}e^{-2j\phi} + t_{12}t_{21}r_{23}r_{21}r_{23}e^{-4j\phi} + \dots$$

Surface 1	Air
	Uniaxial medium
Surface 2	
	Air

4 – PARALLEL AND PERPENDICULAR COUPLING

As you saw before, our equations have parallel and perpendicular field components in them. Those parallel and perpendicular properties can exist in a coupled or a decoupled condition; the former is called, **Oblique Incidence**, and the latter is called, **Normal Incidence**.

- **Oblique Incidence**

For anisotropical media the parallel and perpendicular components of the field vectors amongst the different layers are coupled and no longer independent from each other (as were the case in the former illustrations of the three-layered isotropic mediums). Thus, we will be having four different reflection and transmission coefficients each time the signal penetrates or bounces off the interface.

- **Normal Incidence**

In the normal incidence case (as being handled in this work), we have no coupling between the parallel and perpendicular components. (You can find the proof for this, as I said, in my next book or simply refer to, " *Optical Waves in Layered Media* " by Pochi Yeh).

The difference between **Time** and **Frequency Domains** can be illustrated in the following figures.

For Time Domain

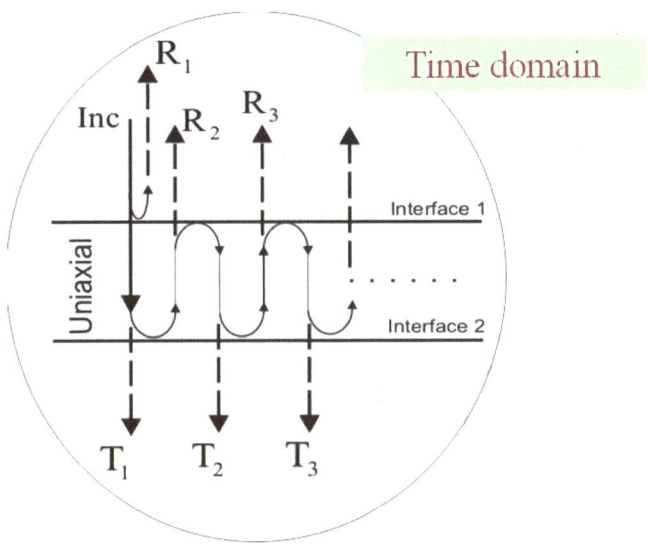

Here, we have recurrent mathematical reflections and transmissions inside the plate itself, and I say "mathematical", because we cannot get rid of them in **Time Domain** analysis. They exist of course physically, but for us to be able to solve our equations easily, we certainly need another method, and this is when **Frequency Domain** analysis comes into play.

For Frequency Domain

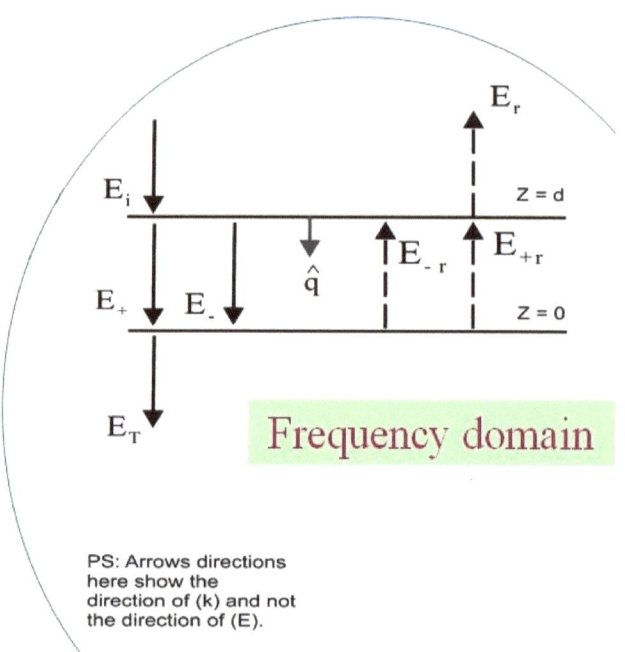

$$E_r$$

$$E_i \qquad Z = d$$

$$\hat{q} \qquad E_{-r} \quad E_{+r}$$

$$E_+ \quad E_- \qquad Z = 0$$

$$E_T \qquad \text{Frequency domain}$$

PS: Arrows directions here show the direction of (k) and not the direction of (E).

Here, we are able to remove the recurrent vibrant waves inside the material theoretically. Such useful mathematical model is called, **Fourier analysis**. **Fourier analysis** of a periodic function refers to the extraction of the series of sines and cosines which, when superimposed, will reproduce the function. This analysis can be expressed as a **Fourier series**. The **Fast Fourier Transform** (**FFT**) is a mathematical method for transforming a function of time into a function of frequency. Here we describe it as transforming from the **Time Domain** to the **Frequency Domain**. I will use Matlab for implementing **FFT** on my chosen pulse, the RC_2. It is very useful for analysis of time-dependent

Simulating Electromagnetic Wave Propagation Through Anisotropic
Uniaxial Plate - For Normal Incidence With Coordinate-Free
Approach

phenomena, such as travelling Electromagnetic Waves.

Other useful application for the **Fourier Transform** is
that when we know all frequency and phase
information about a wave then we may well reconstruct
the original wave precisely, so it is possible to recover a
function from its **Fourier transform**.

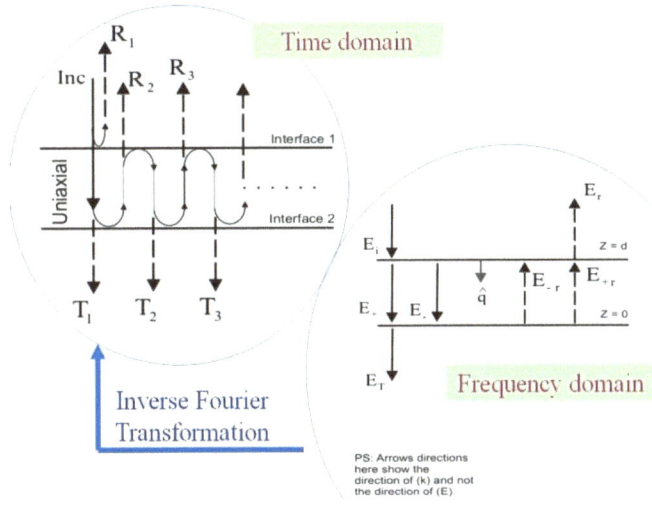

5 - WAVE MATRIX OF A UNIAXIAL MEDIUM

The lack of rotational symmetry in the atomic arrangement of a material is a characteristic for crystals. A crystal is a solid material whose atoms are arranged in a highly ordered microscopic structure, forming a crystal lattice that extends in all directions. A material where the atoms have no periodic structure is therefore not a crystal, like, glass; it is called instead, amorphous solid. Therefore, the phase velocity in crystals depends on the direction of wave propagation.

For an orthogonal coordinate system, the dielectric tensor of a uniaxial medium takes the matrix form:

$$\overline{\varepsilon} = \begin{bmatrix} \varepsilon_{\parallel} & 0 & 0 \\ 0 & \varepsilon_{\perp} & 0 \\ 0 & 0 & \varepsilon_{\perp} \end{bmatrix} \qquad \boxed{\textbf{Matrix form}}$$

Which is a real symmetric matrix, and the dielectric tensor is frequency independent (as chosen in this work) .

$$\overline{\varepsilon} = \varepsilon_{\perp}\overline{I} + (\varepsilon_{\parallel} - \varepsilon_{\perp})\hat{c}\hat{c} \qquad \boxed{\textbf{Dyadic form}}$$

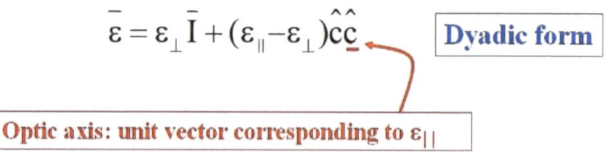

$\boxed{\textbf{Optic axis: unit vector corresponding to } \varepsilon_{\parallel}}$

From the Matrix Form you can see that the parallel and perpendicular components are decoupled, hence, the symmetry of the matrix. You can also notice that I am not using x, y or z directions at all, because this thesis is concerned with the coordinate-free approach instead. For this reason we use an axis we call, optic aixs (i.e., $\hat{c}$), that corresponds to the non-repeated eigenvalue of ε_{II} . Note that, $\hat{c}$ is a unit eigenvector of the dielectric tensor, $\overline{\varepsilon}$.

Decoupling Maxwell's equations would give:

$$\left[k_0^2 \overline{\varepsilon} + (k \times \overline{I})^2 \right] \cdot E_0 = \overline{0}$$

Or by substituting the dyadic form of the dielectric tensor in the above equation, we get

$$\left[(k_0^2 \varepsilon_\perp - k^2)\overline{I} + kk + k_0^2 (\varepsilon_{||} - \varepsilon_\perp)\hat{c}\hat{c} \right] \cdot E_0 = \overline{0}$$

$$\underbrace{}_{\overline{W_u}(k)}$$

→ Wave matrix of uniaxial medium

A non trivial solution of the equation exist by having:

$$\left| \overline{W_u(k)} \right| = 0$$

By some algebraic manipulation, we get

$$\left| \overline{W}_u(k) \right| = k_0^2 (k^2 - k_0^2 \varepsilon_\perp)[(k \cdot \overline{\varepsilon} \cdot k) - k_0^2 \varepsilon_\perp \varepsilon_{||}] = 0$$

which is called as *Dispersion equation*.

13

Surface 1	Air
	Uniaxial medium
Surface 2	
	Air

6 - DIRECTION OF UNIAXIAL MEDIUM'S FIELD VECTORS

$$\text{adj} \ \overline{\overline{W}}_u(k) = (k_0^2 \varepsilon_\perp - k^2)\left[k_0^2 \varepsilon_\| \bar{I} - kk - k_0^2(\varepsilon_\| - \varepsilon_\perp)\hat{c}\hat{c} \right]$$

$$+ k_0^2(\varepsilon_\| - \varepsilon_\perp)(k \times \hat{c})(k \times \hat{c})$$

$$\left[\text{adj} \ \overline{\overline{W}}_u(k) \right].u = E_0 \qquad \text{since:} \quad \boxed{ \overline{\overline{W}}_u(k) \cdot E_0 = \bar{0} }$$

Arbitrary vector

$$\left(\overline{\overline{W}}_u(k) \cdot \underbrace{\left(\overline{\overline{W}}_u(k) \right)^{-1} \left| \overline{\overline{W}}_u(k) \right|}_{\text{adj}\left(\overline{\overline{W}}_u(k) \right)} = \left| \overline{\overline{W}}_u(k) \right| \bar{I} = \bar{0} \right) \cdot u$$

The equation,

$$\left[\text{adj} \ \overline{\overline{W}}_u(k) \right].u = E_0$$

tells us that **u** could be any arbitrary vector, therefore, only the direction of **E₀** is determined and not its magnitude.

The rest of the field vectors can be determined from Maxwell's equations (full detailed derivation will be included in my next book).

Surface 1	Air
	Uniaxial medium
Surface 2	
	Air

7 - WAVE VECTORS AT ISOTROPIC-UNIAXIAL/UNIAXIAL-ISOTROPIC INTERFACES

Dispersion equation

$$\left|\overline{W}_u(k)\right| = k_0^2 \underbrace{(k^2 - k_0^2\varepsilon_\perp)}[(k\cdot\overline{\varepsilon}\cdot k) - k_0^2\varepsilon_\perp\varepsilon_\parallel] = 0$$

$$k_+ = k_0\sqrt{\varepsilon_\perp}$$

Ordinary wave number

$$k_- = k_0\sqrt{\frac{\varepsilon_\perp\varepsilon_\parallel}{\hat{k}_-\cdot\overline{\varepsilon}\cdot\hat{k}_-}}$$

Extraordinary wave number

According to the laws of reflection and refraction, we may represent any of the wave vectors k_i, k_r, k_+, or k_- in general form

$$k_\alpha = b + |q_\alpha|\hat{q}$$

$$b = \hat{q}\times a$$

$$a = k_\alpha\times\hat{q} = b\times\hat{q}$$

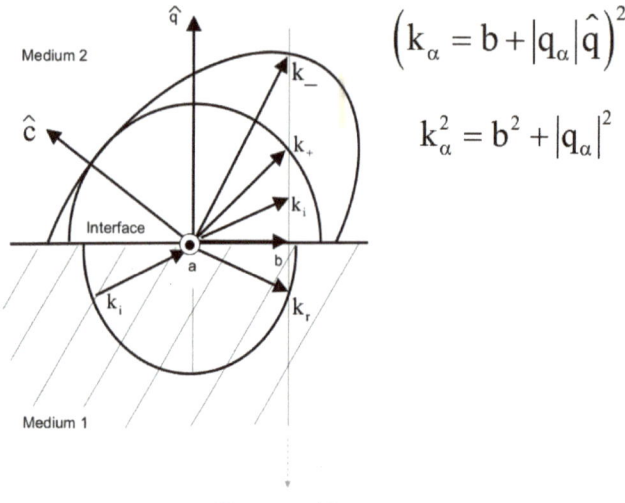

$$\left(k_\alpha = b + |q_\alpha| \hat{q}\right)^2$$

$$k_\alpha^2 = b^2 + |q_\alpha|^2$$

Phase matching

where

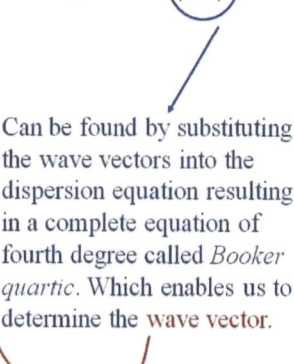

$$\left(k_\alpha = b + |q_\alpha| \hat{q}\right)^2$$

$$k_\alpha^2 = b^2 + |q_\alpha|^2$$

Can be found by substituting the wave vectors into the dispersion equation resulting in a complete equation of fourth degree called *Booker quartic*. Which enables us to determine the wave vector.

8 - REFLECTION & TRANSMISSION AT THE INTERFACE FOR NORMAL INCIDENCE

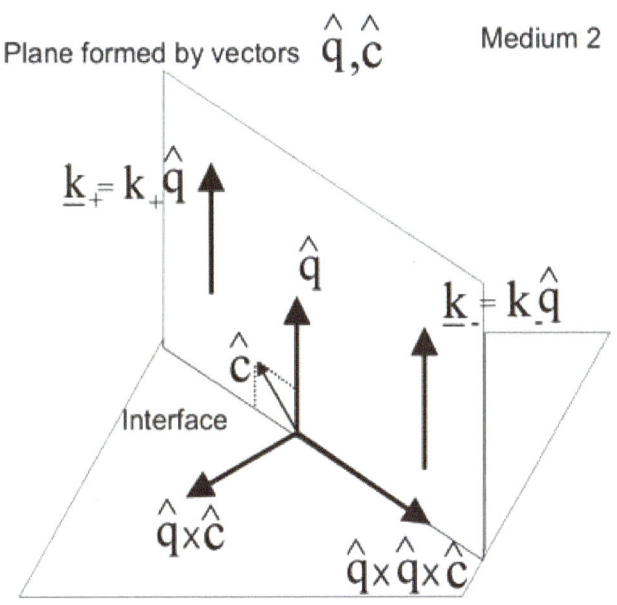

Plane formed by vectors $\hat{q}, \hat{c}$ Medium 2

$\underline{k}_+ = k_+ \hat{q}$

$\hat{q}$

$\underline{k}_- = k_- \hat{q}$

$\hat{c}$

Interface

$\hat{q} \times \hat{c}$

$\hat{q} \times \hat{q} \times \hat{c}$

Incident wave

$$E_{0i} = \left|A_\perp\right|(\hat{q} \times \hat{c}) + \left|A_\parallel\right|\hat{q} \times (\hat{q} \times \hat{c})$$

$$H_{0i} = \sqrt{\frac{\varepsilon_0 \varepsilon_1}{\mu_0}}\left[\left|A_\perp\right|\hat{q} \times (\hat{q} \times \hat{c}) - \left|A_\parallel\right|(\hat{q} \times \hat{c})\right]$$

Reflected wave

$$E_{0r} = \left|B_\perp\right|(\hat{q} \times \hat{c}) + \left|B_\parallel\right|\hat{q} \times (\hat{q} \times \hat{c})$$

$$H_{0r} = \sqrt{\frac{\varepsilon_0\varepsilon_1}{\mu_0}}\left[-\left|B_\perp\right|\hat{q} \times (\hat{q} \times \hat{c}) + \left|B_\parallel\right|(\hat{q} \times \hat{c})\right]$$

Transmitted wave

1. Ordinary

$$E_{0+} = \left|C_+\right|(\hat{q} \times \hat{c})$$

$$H_{0+} = \left|C_+\right|\sqrt{\frac{\varepsilon_0\varepsilon_\perp}{\mu_0}}\left[\hat{q} \times (\hat{q} \times \hat{c})\right]$$

2. Extraordinary

$$E_{0-} = \left|C_-\right|\left[\varepsilon_\perp\hat{c} - \frac{k_-^2}{k_0^2}(\hat{q} \cdot \hat{c})\hat{q}\right]$$

$$H_{0-} = \frac{\varepsilon_\perp\left|k_-\right|}{\omega\mu_0}(\hat{q} \times \hat{c})\left|C_-\right|$$

Simulating Electromagnetic Wave Propagation Through Anisotropic Uniaxial Plate - For Normal Incidence With Coordinate-Free Approach

The boundary conditions at the interface are described by the matrices:

$$\begin{bmatrix} C_+ \\ C_- \end{bmatrix} = \begin{bmatrix} T_{11} & T_{12} \\ T_{21} & T_{22} \end{bmatrix} \begin{bmatrix} A_\perp \\ A_\parallel \end{bmatrix}$$

$$\begin{bmatrix} B_\perp \\ B_\parallel \end{bmatrix} = \begin{bmatrix} \Gamma_{11} & \Gamma_{12} \\ \Gamma_{21} & \Gamma_{22} \end{bmatrix} \begin{bmatrix} A_\perp \\ A_\parallel \end{bmatrix}$$

Surface 1 ———————————————— Air

Uniaxial
medium

Surface 2 ————————————————

Air

9 - Reflection & Transmission at the Plate for Normal Incidence

a) Optic axis perpendicular to the normal of the interface.

$$\text{Term1} = -\ e^{-jk\_d} + \frac{1 - \sqrt{\dfrac{\varepsilon_{\parallel}}{\varepsilon_1}}}{1 + \sqrt{\dfrac{\varepsilon_{\parallel}}{\varepsilon_1}}}\ e^{jk\_d}$$

$$\text{Term2} = -\sqrt{\dfrac{\varepsilon_{\parallel}}{\varepsilon_1}}\ e^{-jk\_d} - \frac{\sqrt{\dfrac{\varepsilon_{\parallel}}{\varepsilon_1}} - \dfrac{\varepsilon_{\parallel}}{\varepsilon_1}}{1 + \sqrt{\dfrac{\varepsilon_{\parallel}}{\varepsilon_1}}}\ e^{jk\_d}$$

$$\Gamma_{22} = \frac{(\text{Term1} - \text{Term2})}{(\text{Term1} + \text{Term2})}$$

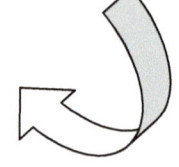

Simulating Electromagnetic Wave Propagation Through Anisotropic Uniaxial Plate - For Normal Incidence With Coordinate-Free Approach

$$\text{Term1} = -\sqrt{\frac{\varepsilon_\perp}{\varepsilon_1}}\, e^{-jk_+d} \;-\; \sqrt{\frac{\varepsilon_\perp}{\varepsilon_1}} \;*\; \frac{1-\sqrt{\dfrac{\varepsilon_\perp}{\varepsilon_1}}}{1+\sqrt{\dfrac{\varepsilon_\perp}{\varepsilon_1}}}\, e^{jk_+d}$$

$$\text{Term2} = -\, e^{-jk_+d} \;-\; \frac{1-\sqrt{\dfrac{\varepsilon_\perp}{\varepsilon_1}}}{1+\sqrt{\dfrac{\varepsilon_\perp}{\varepsilon_1}}}\, e^{jk_+d}$$

$$\Gamma_{11} = \frac{\left(\text{Term1} - \text{Term2}\right)}{\left(\text{Term1} + \text{Term2}\right)}$$

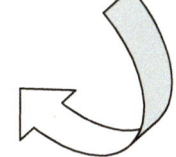

$$T_{11} = \frac{-4}{2j\left(\sqrt{\dfrac{\varepsilon_\perp}{\varepsilon_1}} + \sqrt{\dfrac{\varepsilon_1}{\varepsilon_\perp}}\right)\sin(k_+d) \,-\, 2\cos(k_+d)}$$

$$T_{22} = \frac{\left(\dfrac{4}{\text{Term1}}\right)}{-\left(1-\text{Term2}\right) + \sqrt{\dfrac{\varepsilon_1}{\varepsilon_\parallel}} \;*\; \left(1+\text{Term2}\right)}$$

where

$$\text{Term1} = \sqrt{\frac{\varepsilon_\parallel}{\varepsilon_1}} \ * \ e^{jk\_d} \ - \ e^{jk\_d}$$

$$\text{Term2} = \frac{\sqrt{\dfrac{\varepsilon_\parallel}{\varepsilon_1}} \ * \ e^{-jk\_d} \ + \ e^{-jk\_d}}{\sqrt{\dfrac{\varepsilon_\parallel}{\varepsilon_1}} \ * \ e^{jk\_d} \ - \ e^{jk\_d}}$$

$$k_i = k_0 \sqrt{\varepsilon_1} = -k_r$$

$$k_+ = k_0 \sqrt{\varepsilon_\perp}$$

$$k_- = k_0 \sqrt{\frac{\varepsilon_\perp \varepsilon_\parallel}{\hat{q} \cdot \overline{\overline{\varepsilon}} \cdot \hat{q}}} = k_0 \sqrt{\varepsilon_\parallel}$$

Simulating Electromagnetic Wave Propagation Through Anisotropic
Uniaxial Plate - For Normal Incidence With Coordinate-Free
Approach

$$\begin{bmatrix} F_\perp \\ F_\parallel \end{bmatrix} = \begin{bmatrix} T_{11}e^{jk_0d} & 0 \\ 0 & T_{22}e^{jk_0d} \end{bmatrix} \begin{bmatrix} A_\perp \\ A_\parallel \end{bmatrix}$$

$$\begin{bmatrix} B_\perp \\ B_\parallel \end{bmatrix} = \begin{bmatrix} \Gamma_{11}e^{2jk_0d} & 0 \\ 0 & \Gamma_{22}e^{2jk_0d} \end{bmatrix} \begin{bmatrix} A_\perp \\ A_\parallel \end{bmatrix}$$

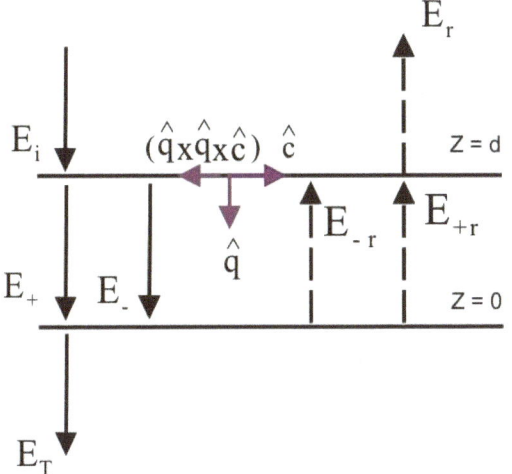

PS: Arrows directions
here show the
direction of (k) and not
the direction of (E).

$$E_+ = C_+ \ (\hat{q} \times \hat{c}) \ e^{-jk_+ z}$$

$$E_{+r} = D_+ \ (\hat{q} \times \hat{c}) \ e^{jk_+ z}$$

$$E_- = \varepsilon_\perp \ C_- \ \hat{c} \ e^{-jk_- z}$$

$$E_{-r} = \varepsilon_\perp \ D_- \ \hat{c} \ e^{jk_- z}$$

$$E_i = |A_\perp| \ (\hat{q} \times \hat{c}) \ e^{-jk_0 z} + |A_\parallel| \ \hat{q} \times (\hat{q} \times \hat{c}) \ e^{-jk_0 z}$$

$$E_r = |B_\perp| \ (\hat{q} \times \hat{c}) \ e^{jk_0 z} + |B_\parallel| \ \hat{q} \times (\hat{q} \times \hat{c}) \ e^{jk_0 z}$$

$$E_T = |F_\perp| \ (\hat{q} \times \hat{c}) \ e^{-jk_0 z} + |F_\parallel| \ \hat{q} \times (\hat{q} \times \hat{c}) \ e^{-jk_0 z}$$

I used for this simulation: Thickness of plate=0.2 m, f=1 GHz.

Calculating for:

$$\Gamma_{22} \quad T_{22}$$

$$V_{p-} = w \, / \, k_-$$

delivers the following results:

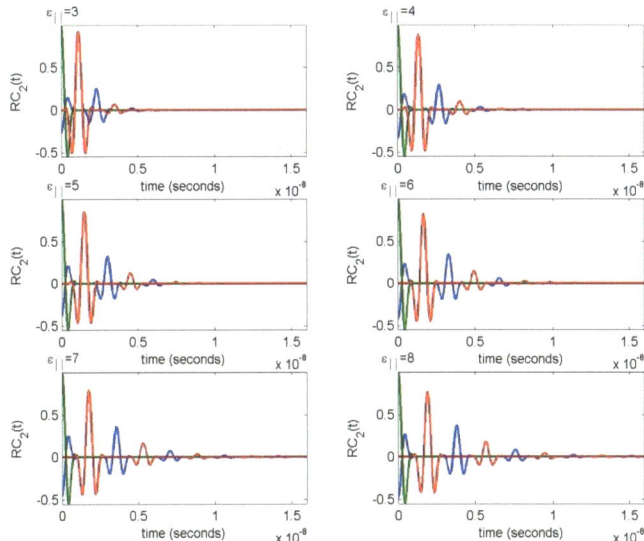

And

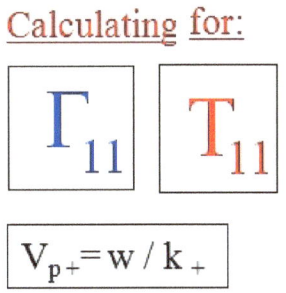

delivers the following results:

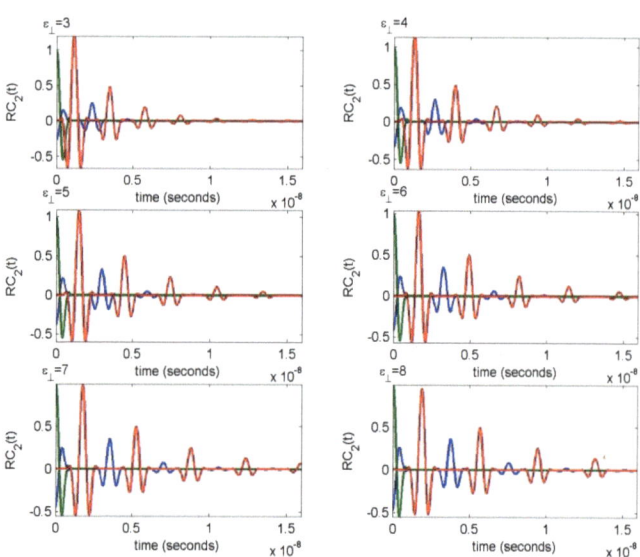

b) Optic axis parallel to the normal of the interface.

$$k_t = k_+ = k_- = |k_0| \sqrt{\varepsilon_\perp}\ \hat{c}$$

This is the case where both wave vectors have the same speed and direction as if the second medium were isotropic.

Simulating Electromagnetic Wave Propagation Through Anisotropic Uniaxial Plate - For Normal Incidence With Coordinate-Free Approach

<u>where</u>

$$k_t \cdot \boxed{D_{0t}} = 0$$

<u>implies</u>

$$\hat{c} \cdot E_{0t} = 0$$

Electric flux density.

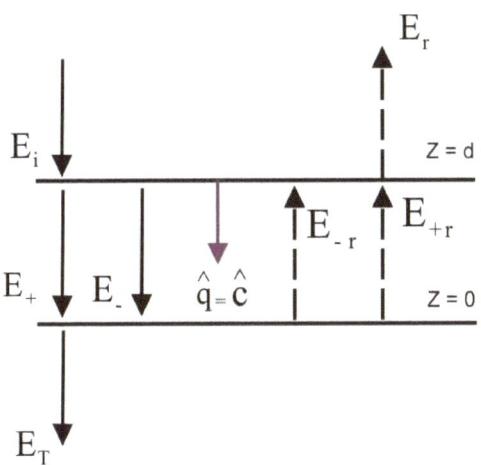

PS: Arrows directions
here show the
direction of (k) and not
the direction of (E).

c) Optic axis having an angle of 45° with the normal of the interface.

$$k_i = k_0 \sqrt{\varepsilon_1} = -k_r$$

$$k_+ = k_0 \sqrt{\varepsilon_\perp}$$

$$k_- = k_0 \sqrt{\frac{\varepsilon_\perp \varepsilon_\|}{\hat{q} \cdot \bar{\bar{\varepsilon}} \cdot \hat{q}}} = k_0 \sqrt{\frac{\varepsilon_\perp \varepsilon_\|}{\varepsilon_\perp + \left(\varepsilon_\| - \varepsilon_\perp\right) * \cos^2(\beta)}}$$

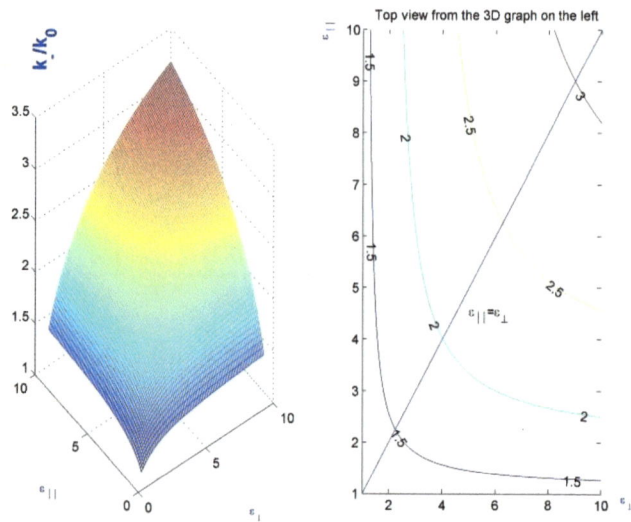

Simulating Electromagnetic Wave Propagation Through Anisotropic Uniaxial Plate - For Normal Incidence With Coordinate-Free Approach

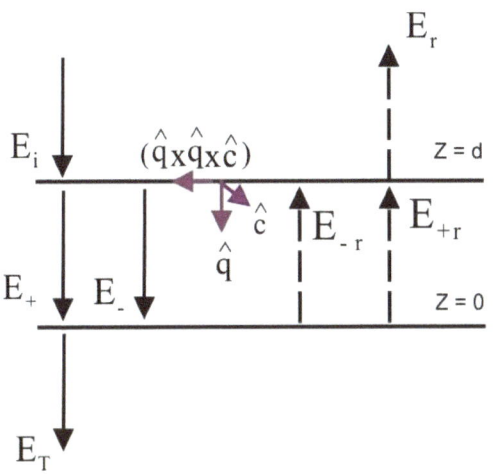

PS: Arrows directions here show the direction of (k) and not the direction of (E).

$$E_+ = C_+ \left(\hat{q} \times \hat{c} \right) e^{-jk_+z}$$

$$E_{+r} = D_+ \left(\hat{q} \times \hat{c} \right) e^{jk_+z}$$

$$E_- = \left[\varepsilon_\perp \hat{c} - \frac{k_-^2}{k_0^2} \cos(\beta) \hat{q} \right] C_- e^{-jk_-z}$$

$$E_{-r} = \left[\varepsilon_\perp \hat{c} - \frac{k_-^2}{k_0^2} \cos(\beta) \hat{q} \right] D_- e^{jk_-z}$$

$$E_i = \left|A_\perp\right| \, (\hat{q} \times \hat{c}) \, e^{-jk_0 z} + \left|A_\parallel\right| \, \hat{q} \times (\hat{q} \times \hat{c}) \, e^{-jk_0 z}$$

$$E_r = \left|B_\perp\right| \, (\hat{q} \times \hat{c}) \, e^{jk_0 z} + \left|B_\parallel\right| \, \hat{q} \times (\hat{q} \times \hat{c}) \, e^{jk_0 z}$$

$$E_T = \left|F_\perp\right| \, (\hat{q} \times \hat{c}) \, e^{-jk_0 z} + \left|F_\parallel\right| \, \hat{q} \times (\hat{q} \times \hat{c}) \, e^{-jk_0 z}$$

$$\text{Term1} = -e^{-jk\_d} + \frac{1 - \dfrac{k_-}{k_0} \sqrt{\dfrac{1}{\varepsilon_1}}}{1 + \dfrac{k_-}{k_0} \sqrt{\dfrac{1}{\varepsilon_1}}} \, e^{jk\_d}$$

$$\text{Term2} = -\frac{k_-}{k_0} \sqrt{\dfrac{1}{\varepsilon_1}} \, e^{-jk\_d} - \frac{k_-}{k_0} \frac{\sqrt{\dfrac{1}{\varepsilon_1}} - \dfrac{k_-}{\varepsilon_1 * k_0}}{1 + \dfrac{k_-}{k_0} \sqrt{\dfrac{1}{\varepsilon_1}}} \, e^{jk\_d}$$

$$\Gamma_{22} = \frac{\text{Term1} - \text{Term2}}{\text{Term1} + \text{Term2}}$$

Simulating Electromagnetic Wave Propagation Through Anisotropic Uniaxial Plate - For Normal Incidence With Coordinate-Free Approach

$$T_{22} = \frac{\left(\dfrac{4}{\text{Term1}}\right) * \dfrac{k_-}{k_0}}{-\left(1 - \text{Term2}\right) * \dfrac{k_-}{k_0} + \sqrt{\varepsilon_1} * \left(1 + \text{Term2}\right)}$$

where

$$\text{Term1} = \frac{k_-}{k_0}\sqrt{\frac{1}{\varepsilon_1}}\; e^{jk\_d} - e^{jk\_d}$$

$$\text{Term2} = \frac{\dfrac{k_-}{k_0}\sqrt{\dfrac{1}{\varepsilon_1}}\; e^{-jk\_d} + e^{-jk\_d}}{\dfrac{k_-}{k_0}\sqrt{\dfrac{1}{\varepsilon_1}}\; e^{jk\_d} - e^{jk\_d}}$$

Calculating for:

$$\Gamma_{22} \qquad T_{22}$$

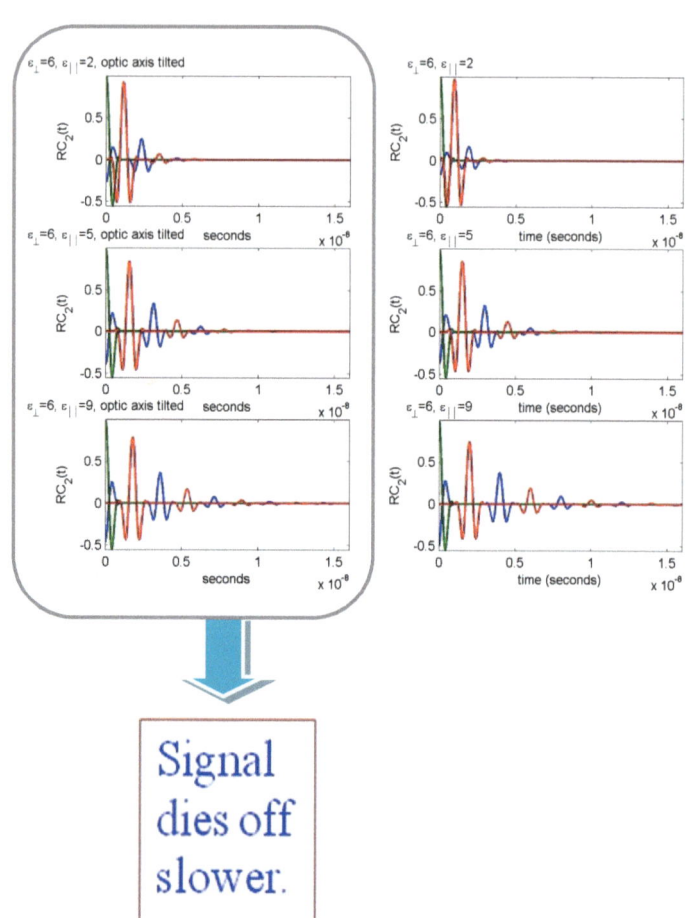

Signal
dies off
slower.

Simulating Electromagnetic Wave Propagation Through Anisotropic Uniaxial Plate - For Normal Incidence With Coordinate-Free Approach

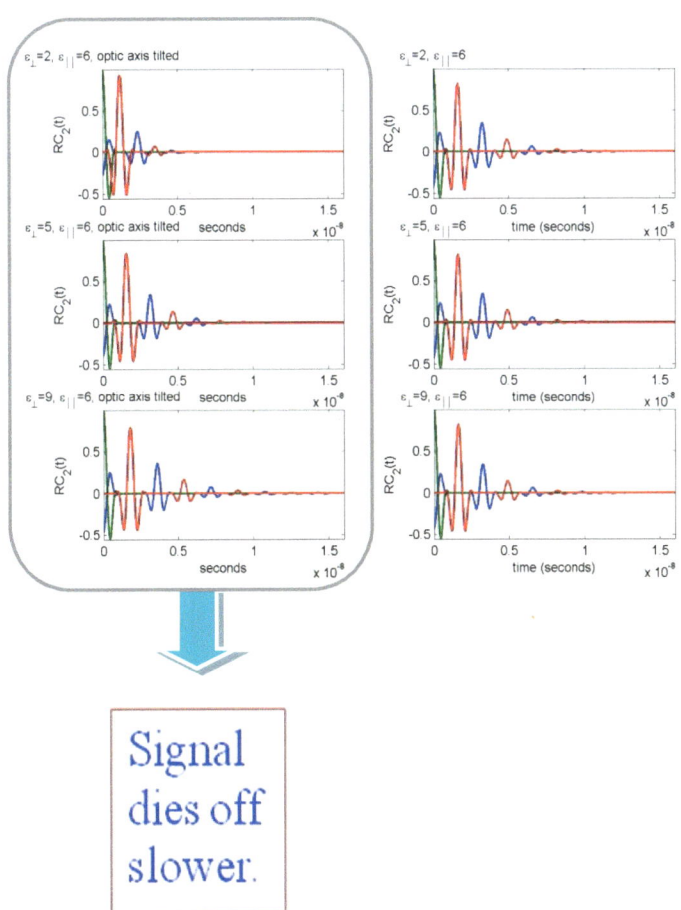

Signal dies off slower.

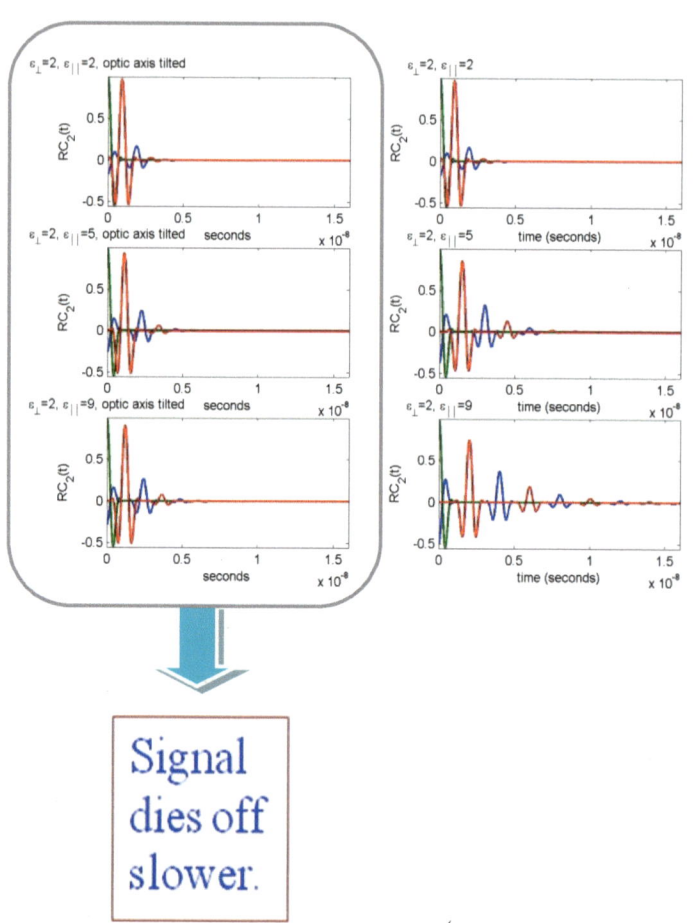

Signal
dies off
slower.

Simulating Electromagnetic Wave Propagation Through Anisotropic Uniaxial Plate - For Normal Incidence With Coordinate-Free Approach

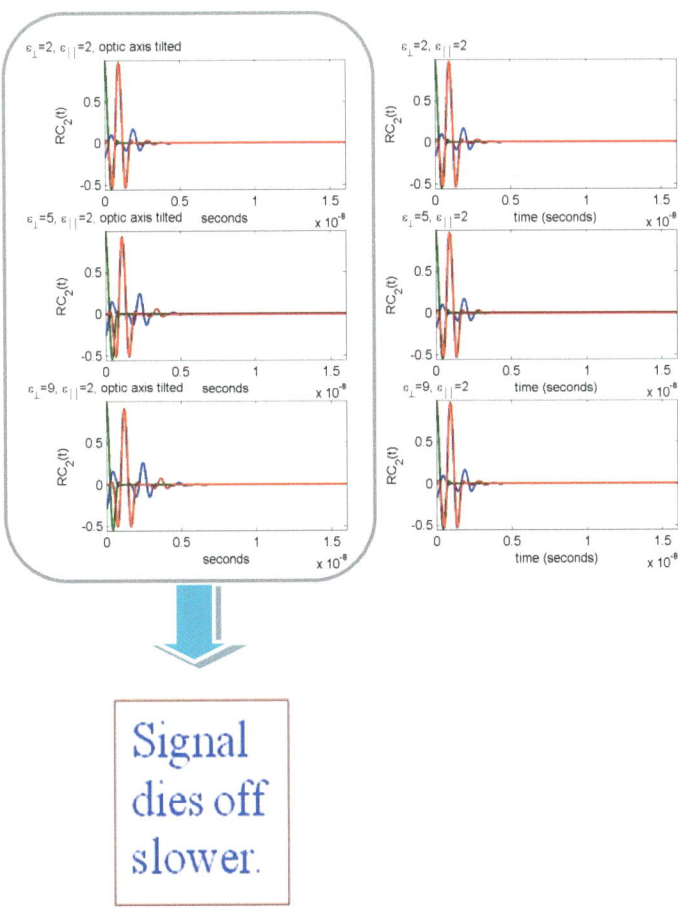

Signal dies off slower.

The first row in the last couple of images is the same and is set with equal permittivity, that is, the perpendicular one equals the parallel one. Notice that they give the exact same output as if this special case were for an isotropic material.

	Air
Surface 1	
	Uniaxial medium
Surface 2	
	Air

10 - APPENDIX

RC₂ Pulse

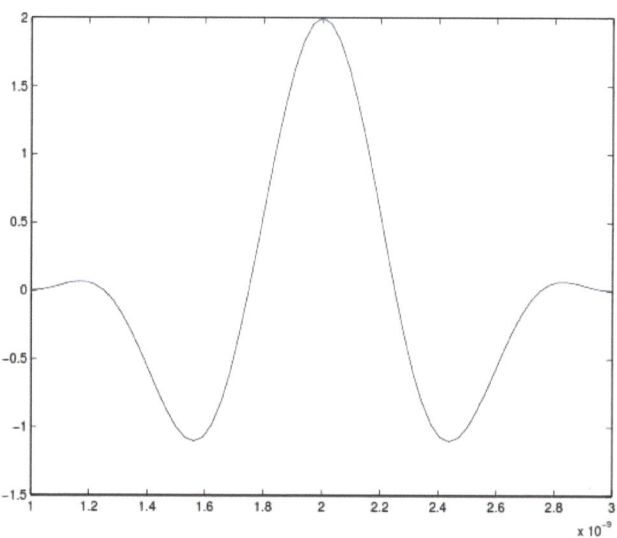

Figure A.1.: RC_2 in Time Domain.

The function of raised cosine pulse of rank 2 is

$$f_{RC_2}(t) = \begin{cases} (1 + \cos(\pi f_0 t)) \cos(2\pi f_0 t) & , -T_0 < t < T_0 \\ \\ 0 & , elsewhere \end{cases}$$

To derive its Frequency Domain final equation we have

36

Simulating Electromagnetic Wave Propagation Through Anisotropic Uniaxial Plate - For Normal Incidence With Coordinate-Free Approach

$$\begin{aligned}
f_{RC_2}(\omega) &= \int_{-T_0}^{T_0} (1 + \frac{e^{j\pi f_0 t}}{2} + \frac{e^{-j\pi f_0 t}}{2})(\frac{e^{2j\pi f_0 t}}{2} + \frac{e^{-2j\pi f_0 t}}{2}).dt \; e^{j\omega t} \\
&= \int_{-T_0}^{T_0} (\frac{e^{(j2\pi f_0 + j\omega)t} + e^{(-j2\pi f_0 + j\omega)t}}{2} + \frac{e^{(j3\pi f_0 + j\omega)t} + e^{(-j\pi f_0 + j\omega)t}}{4} \\
&\quad + \frac{e^{(j\pi f_0 + j\omega)t} + e^{(-j3\pi f_0 + j\omega)t}}{4}).dt \\
&= \sin(\omega T_0)[\frac{2\omega}{-\omega_0^2 + \omega^2}] - \sin(\omega T_0)[\frac{\omega}{-(\frac{3}{2}\omega_0)^2 + \omega^2}] \\
&\quad - \sin(\omega T_0)[\frac{\omega}{-(\frac{\omega_0}{2})^2 + \omega^2}]
\end{aligned}$$

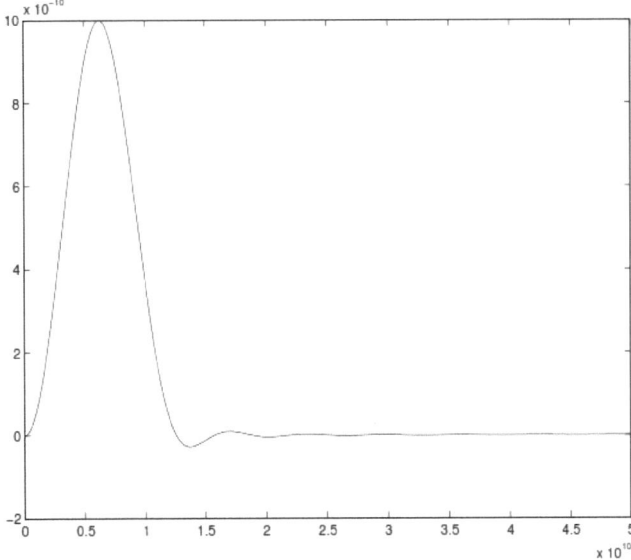

Figure A.2.: RC_2 in Frequency Domain.

ABOUT THE FRONT COVER IMAGE

I took the picture of the Odonata (aka, Dragonfly) in my garden. Its beautifully glittering wings, yet for the most part transparent, demonstrate a very low level of opacity. Typical transparent media such as glasses are isotropic, which means that light behaves the same way no matter which direction it is travelling to. That is why even when the wings of the Odonata are too thin for us to actually observe how light is being transmitted through them, there is still no double refraction taking place. Even rotating the wings around any axis, including an optic axis that one might assume it exist, will not change its optical behavior, hence, no uniaxial properties exist.
This is evident from the different angles at which its wings are tilting while flickering; one can observe the continuity of the background pattern with that of the transmitted through the wings.